To Zoee

Well, hi!! I'm Amelia Gearheart, and I love learning how things work! What kind of things? Well, I love learning about Science, Technology, Engineering, Math and Trades! Or for short, STEMT.

The Gearheart

That's what the Gearheart is all about! It's a symbol for girls who love learning about how things work through STEMT. In this book we're going to learn about how electricity is made and all the STEMT careers that help make it! And, I'll even introduce you to my friends that have jobs in STEMT! It'll be so fun!!

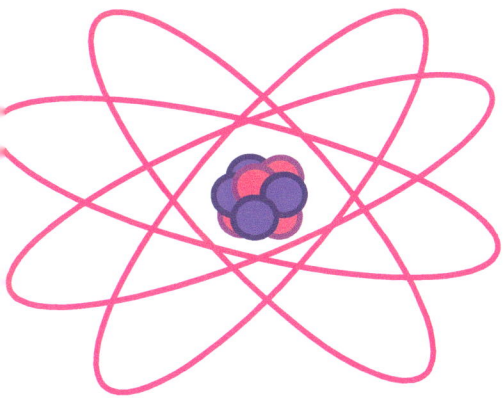

We'll learn about sciences like physics and trades like welding. You'll meet scientists like Beth the Biologist, engineers like Cindy the Civil Engineer, and tradeswomen like Elissa the Electrician.

Each of these ladies have super special jobs that you can have too!

So, keep reading, you'll see! You'll love learning about electricity and STEMT careers, just like I do!

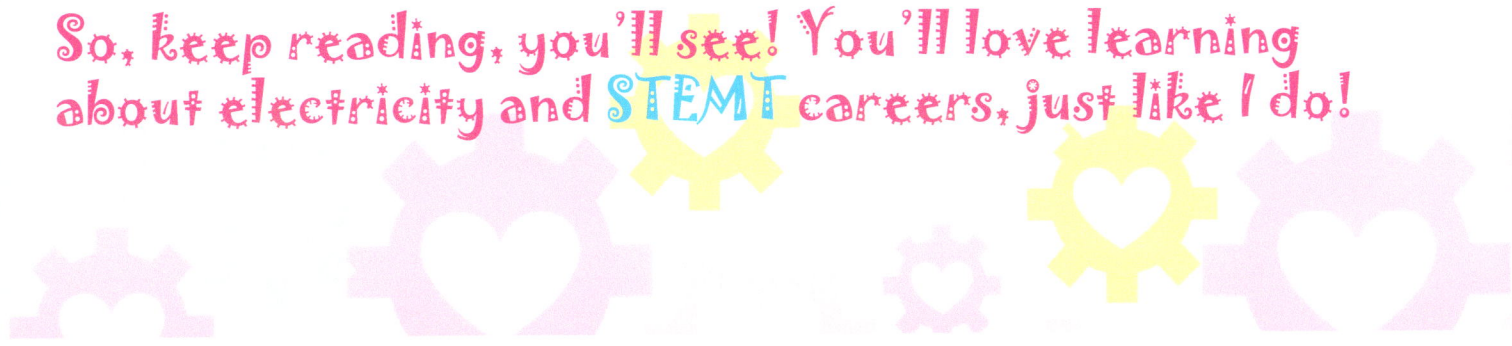

Electricity is everywhere, we use it all the time!

You use electricity when you turn on a computer or lamp or your favorite TV show. So many devices use electricity. Can you name any?

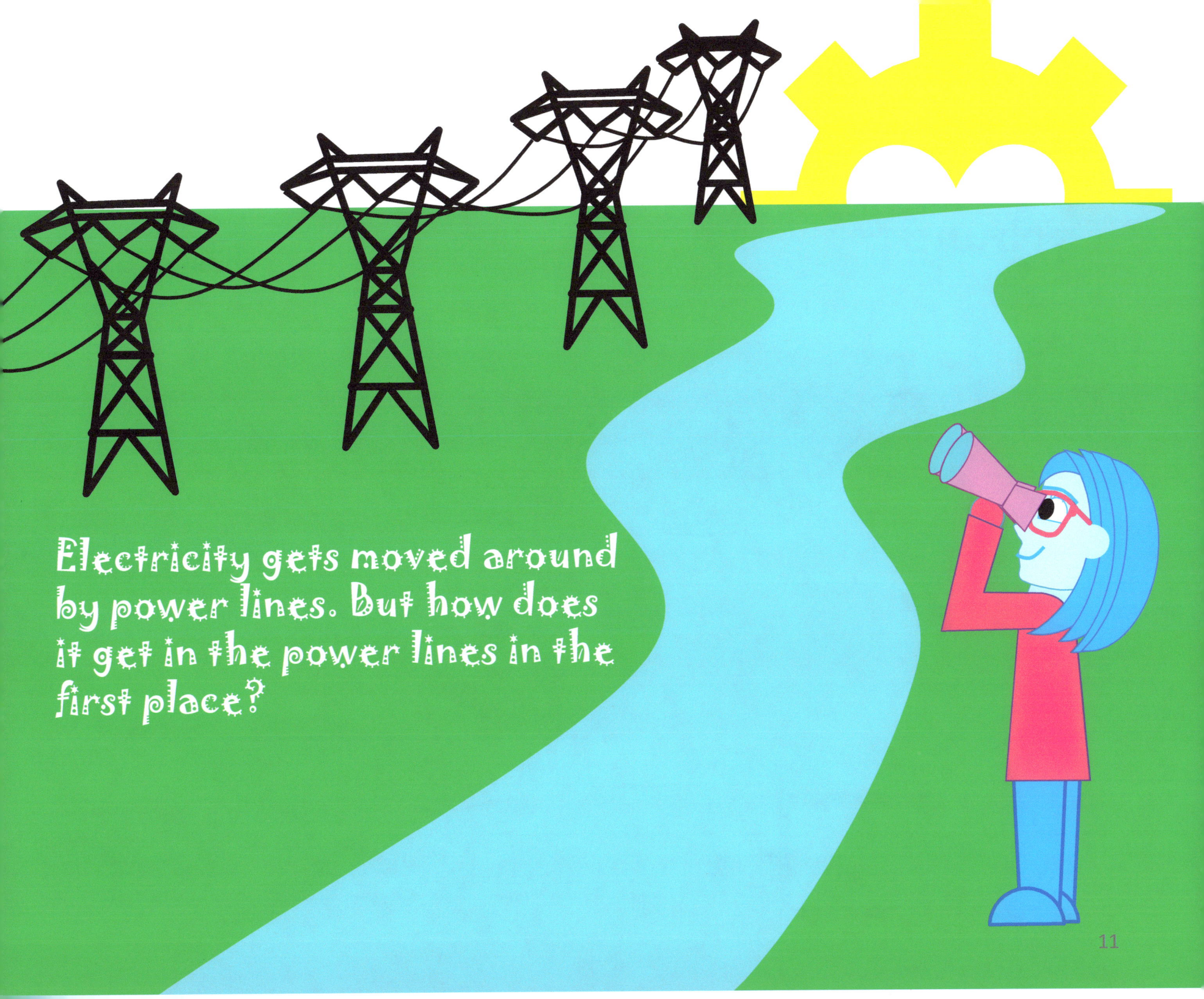

Electricity gets moved around by power lines. But how does it get in the power lines in the first place?

Does electricity just come from the ground?

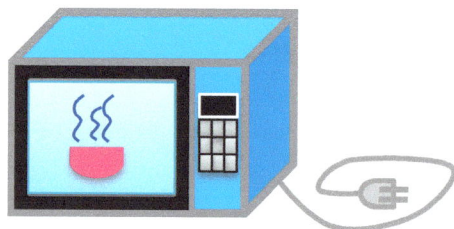

Or Sky?

Or does electricity come from space?

# Well, what is electricity anyway?

The Atom

To understand electricity, we must learn about atoms. Atoms make up everything - plants, rocks, you, and me!

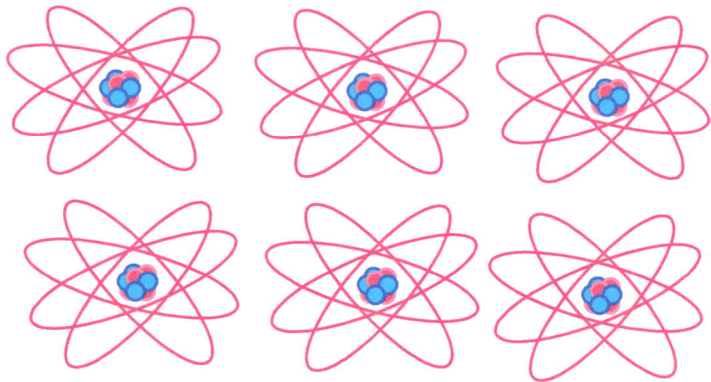

Atoms are WAAAAY too small for your eyes to see. There are different kinds of atoms - gas atoms, metal atoms, and more!

The atom has two parts,
A nucleus and electron cloud.
The nucleus is a heavy part in the
middle, and the electrons orbit
around the nucleus.

Electron

Nucleus

Fahima
The
Physicist

Fahima the Physicist
studies atoms and lots of other
"physical sciences."
She and other scientists
look for new and better ways
to make electricity with electrons.

Metal atoms are special when it comes to electricity. You see, metal atoms share their electrons with each other. This means the electrons can run around freely between metal atoms.

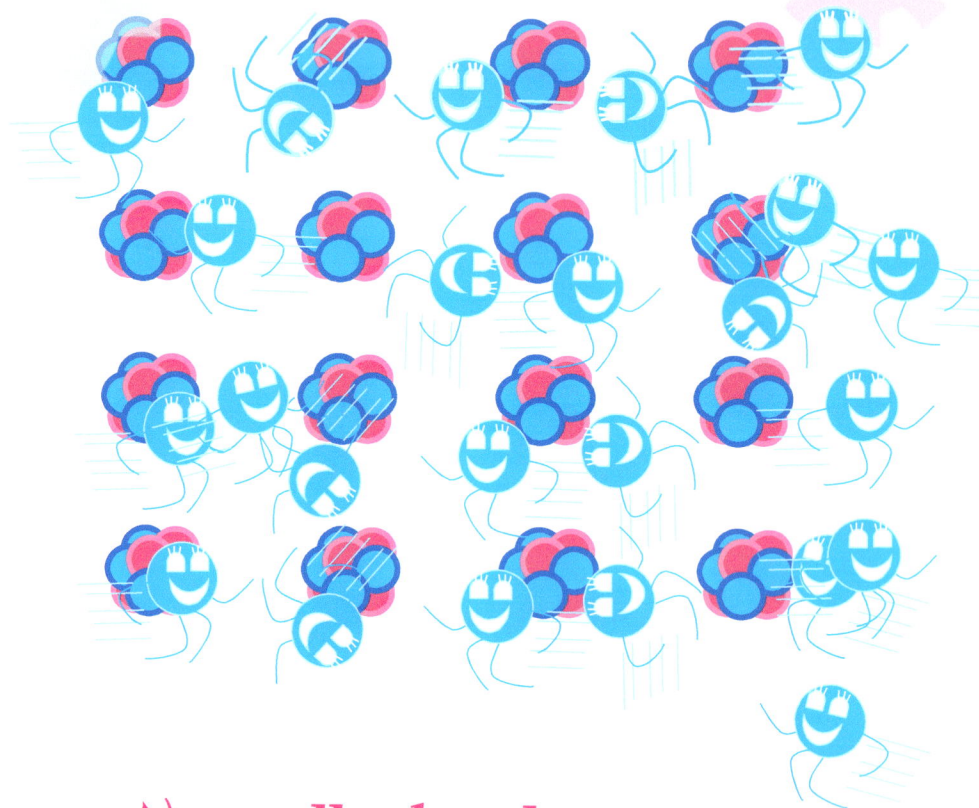

Normally the electrons jump between atoms randomly. But if you can get them to move in one direction, all together, you can make electricity!

But how do you get electrons to move all together? Scientists like Fahima have discovered that electrons are attracted by magnets (magnets are like electrons' favorite superstar!). They love to follow magnets around, like a magnet paparazzi!

So, by moving magnets around metal wire, electrons will move through the wire all together, and electricity is made!

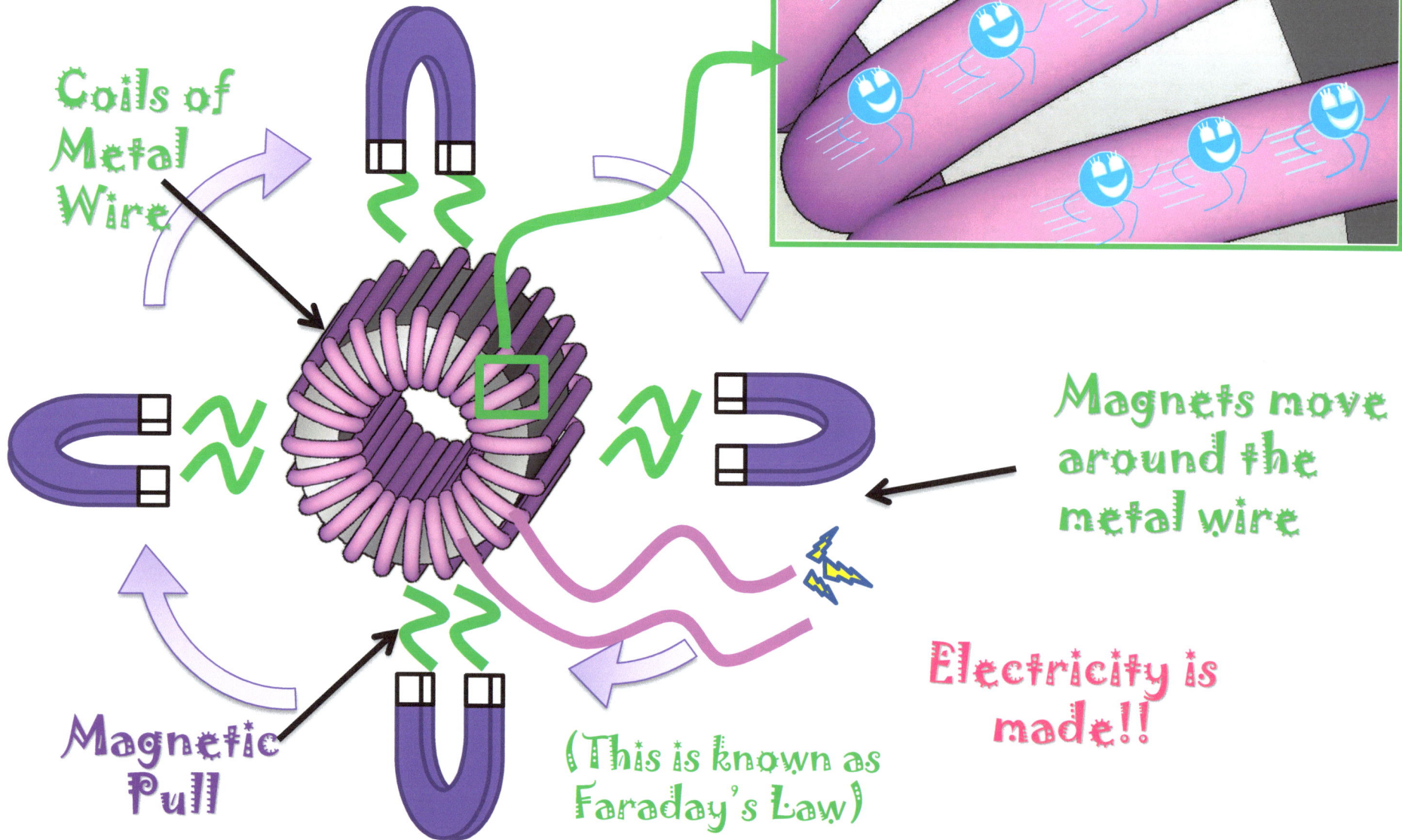

Coils of Metal Wire

Magnets move around the metal wire

Magnetic Pull

(This is known as Faraday's Law)

Electricity is made!!

The electrons don't mind, in fact they'd love to chase magnets all day. But how do you make the magnets move around and around all day long? That's a lot of work!

What's something that goes around and around, and can be pushed by moving gas, steam or air? A fan, of course! When used in reverse, fans can move magnets to make electricity.

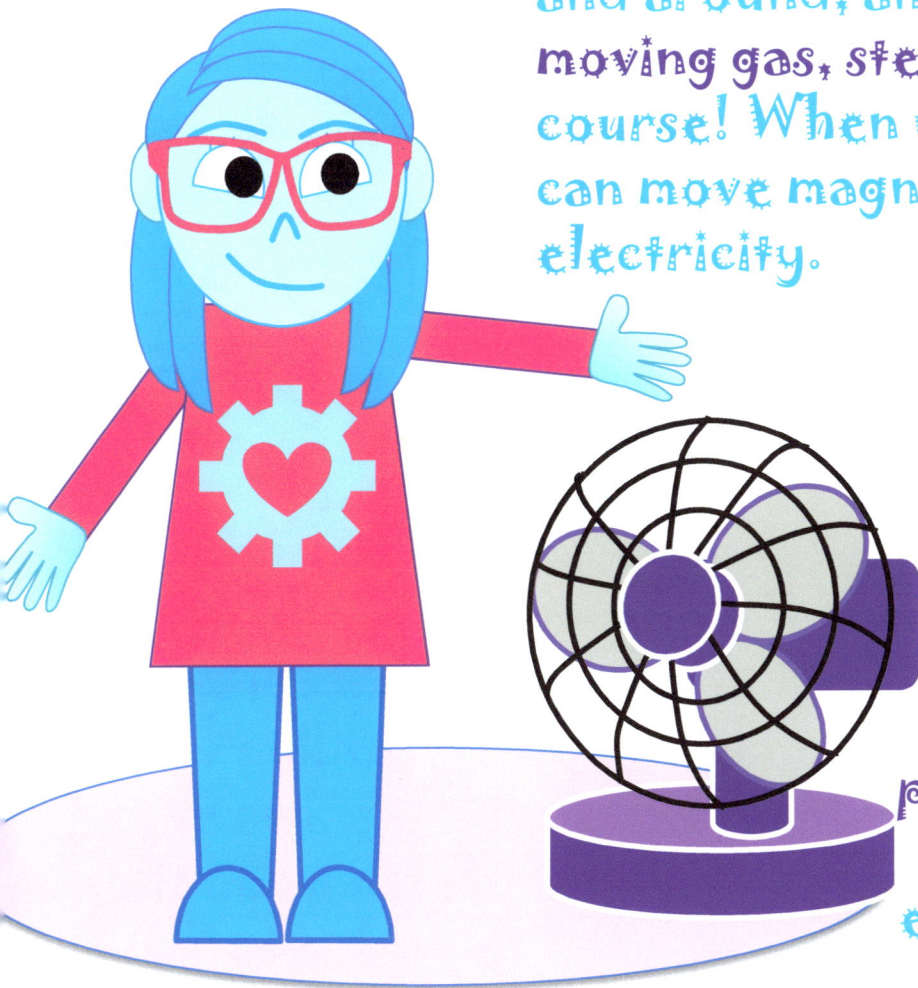

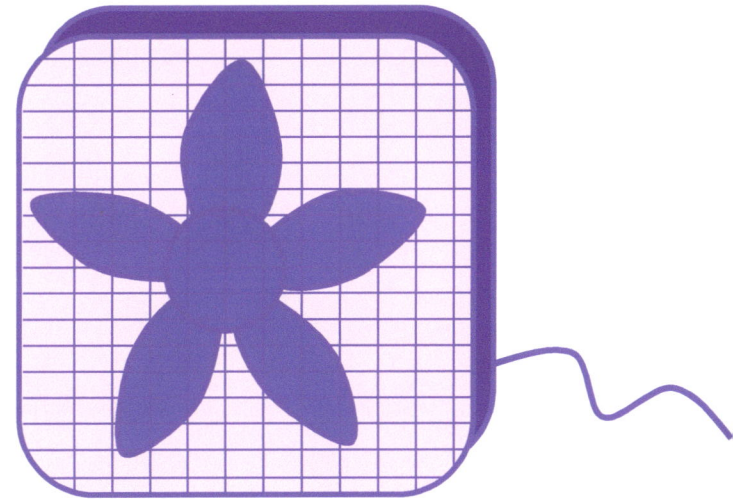

But it's not just any fan that can be pushed by gas, steam, or air to make electricity. Do you know what an electricity-generating fan is called?

A giant metal fan called a turbine is one machine used to make or generate electricity. Turbines come in all shapes and sizes!

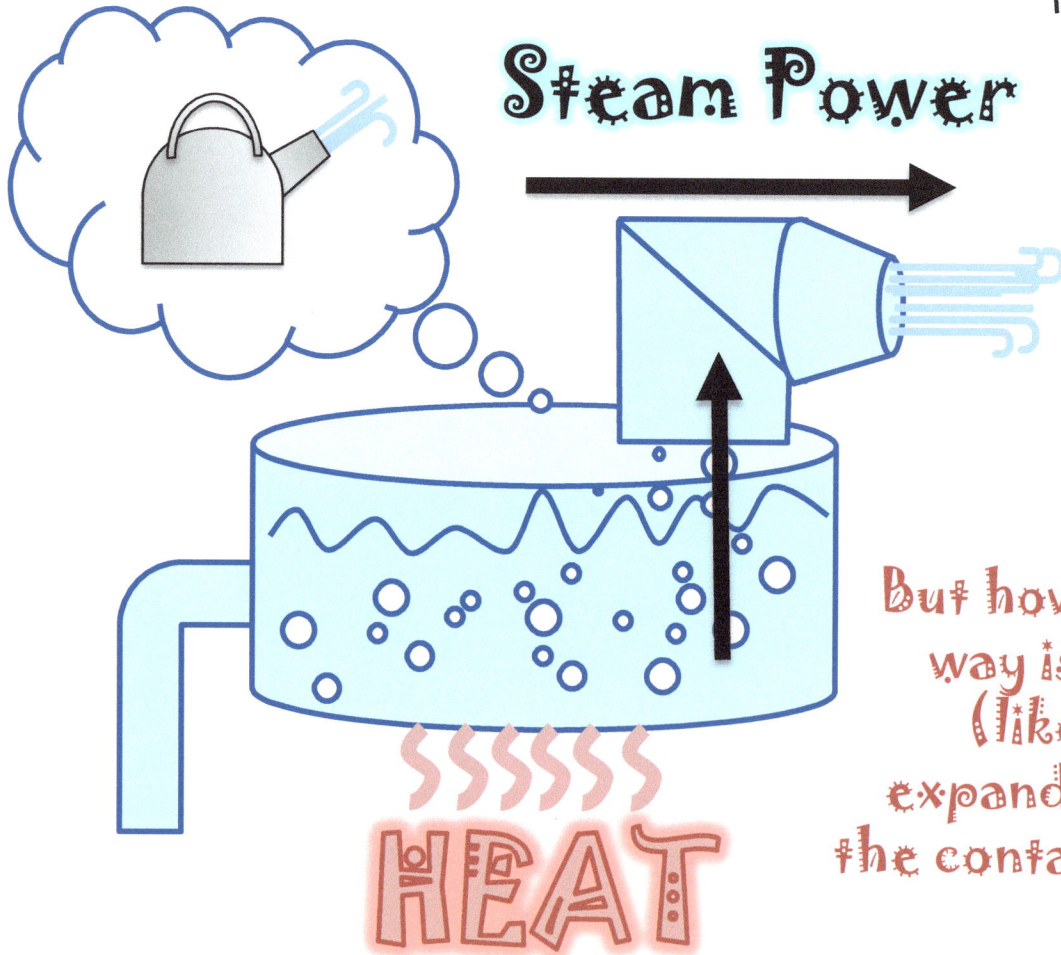

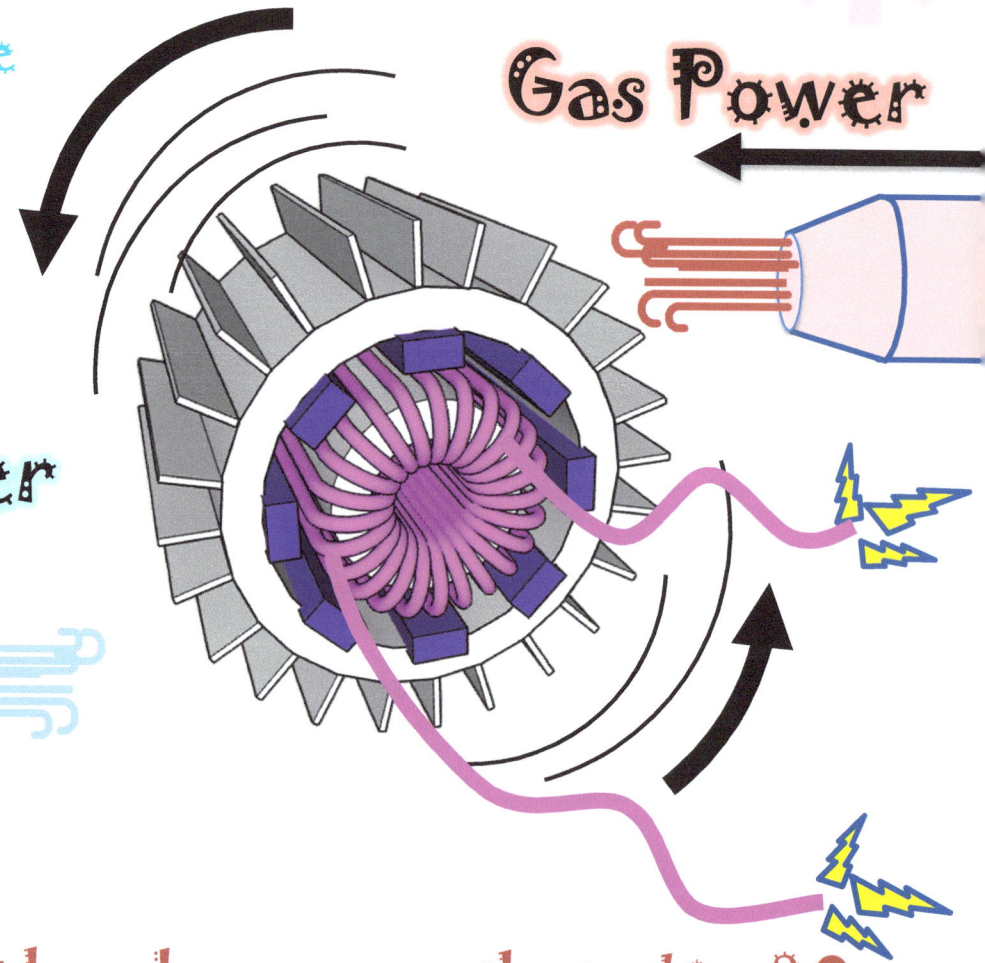

## Gas Power

## Steam Power

**HEAT**

But how do you move the turbine? One way is to boil water in a big container (like a giant teapot!). Boiling water expands into steam, which blasts out of the container! That jet of steam can then be used to push a turbine.

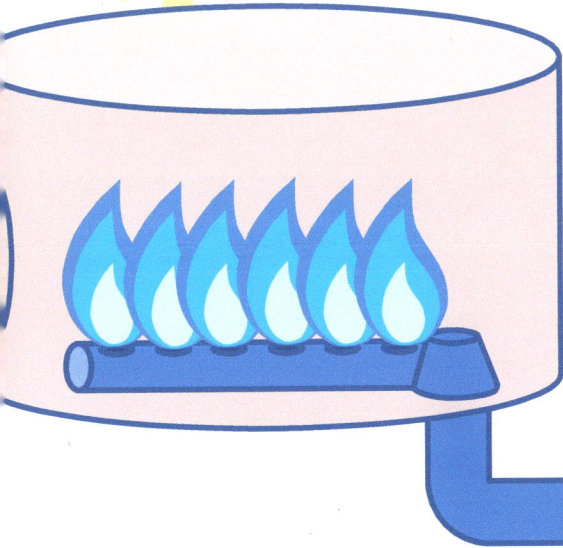

You can also burn liquid fuels to power turbines (like gasoline, kerosene, natural gas, oil and others) because they expand when they burn, too.

**Gas In**

So, boiling water to make steam or burning fuels is what makes most turbines spin. When the turbine spins, magnets move around metal wire, and this makes electricity!

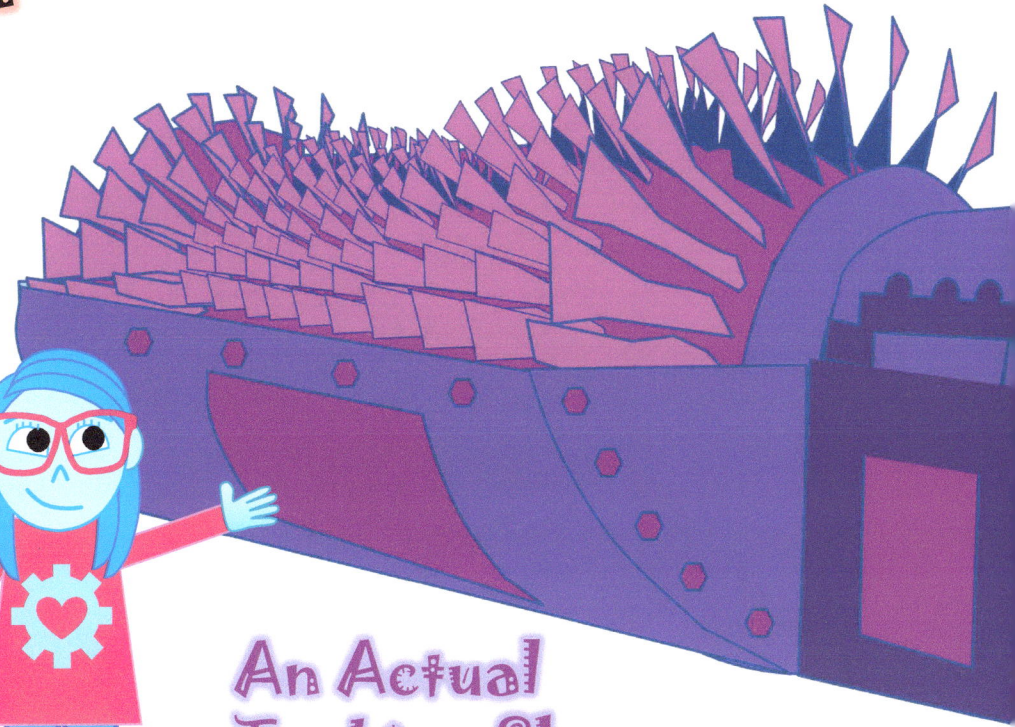

An Actual Turbine Shape

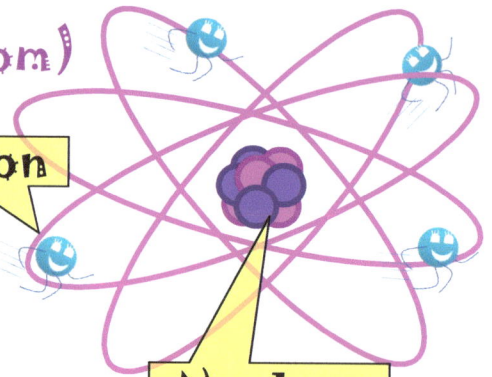

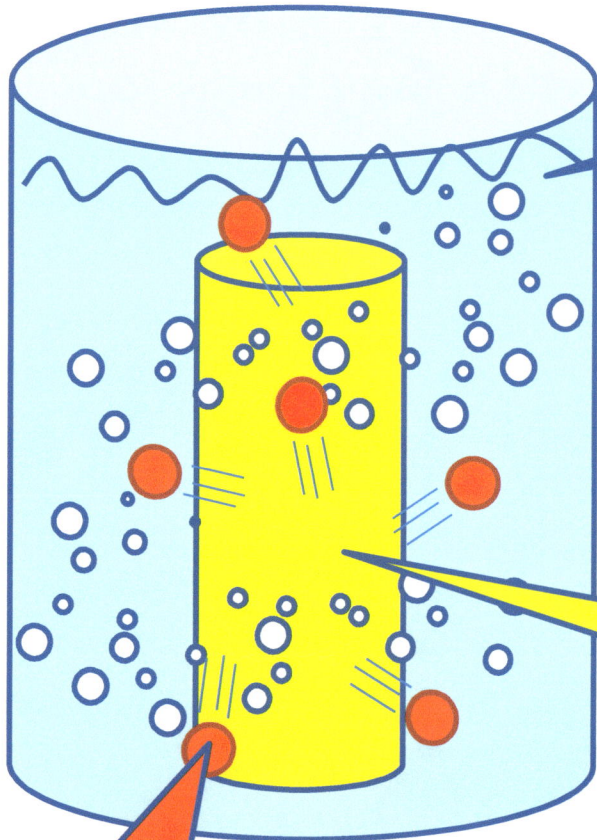

Nuclear metals go into **nuclear reactors** or big facilities designed to harness all the heat from nuclear metals. Nuclear reactors are very powerful! Lots of STEMT careers go into designing and operating nuclear reactor power plants!

## Nuclear Reactor Operating Room

## Nina the Nuclear Engineer
(designs nuclear power plants)

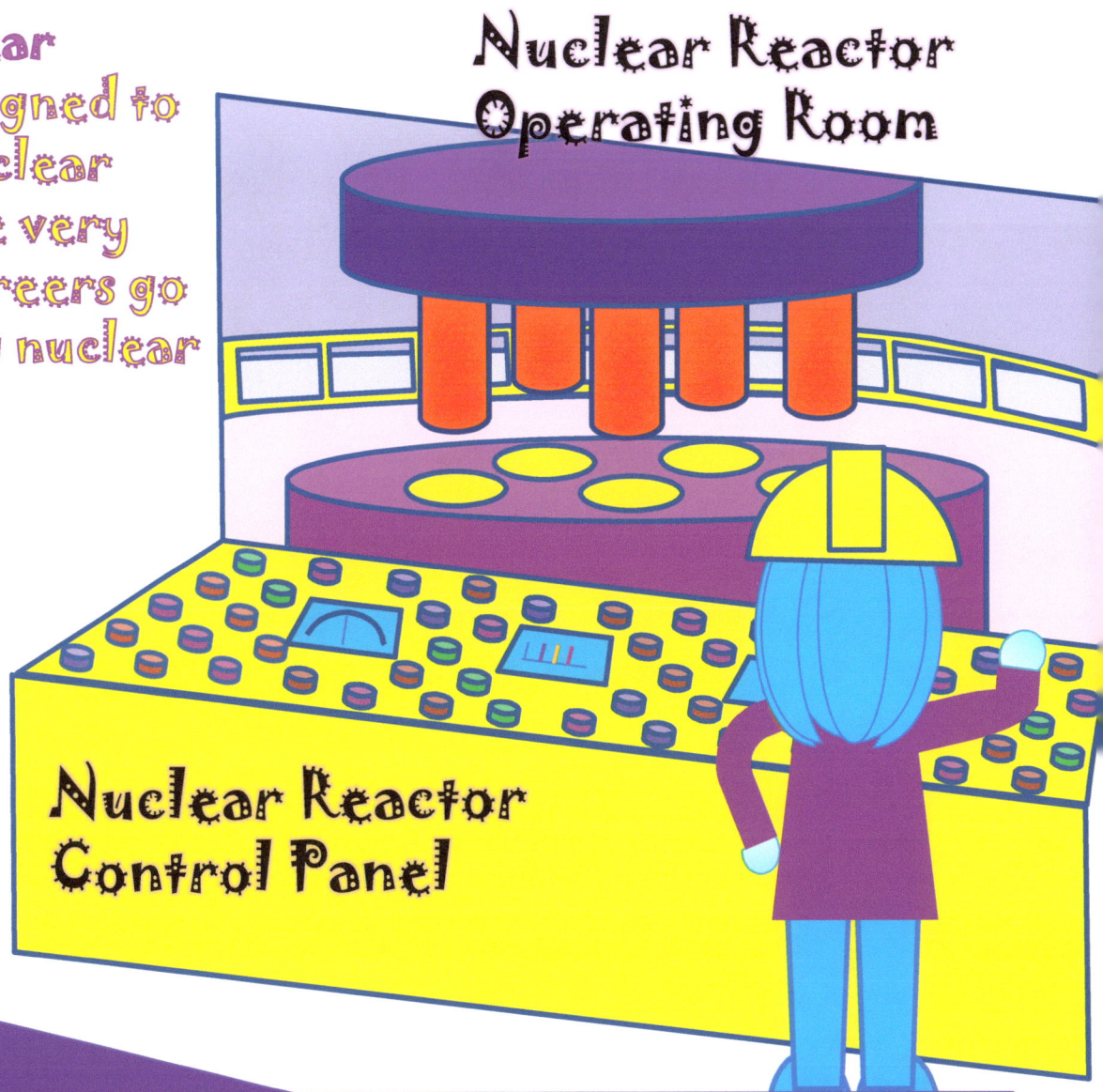

## Nuclear Reactor Control Panel

Hi, I'm Nina! Nuclear Engineers like me work with physicists like Fahima to design better nuclear power plants. Fahima studies nuclear decay while I design the facility. Lots of other engineers, scientists, and trades folks help us, too!

Who builds the nuclear power plants, the turbines and everything else? Trades-folks of course, like Mia, Wendy and Mindy! They love that their jobs are hands-on, meaning, they get to build things all day long!

Mia the Machinist

Wendy the Welder

Mindy the Mechanic

Mia's Milling Machine

Mia is a machinist. She uses big machines like this milling machine to cut metal into useful shapes. Wendy is a welder, which means she joins metals together into different useful shapes! Mindy the mechanic uses tools and fasteners to assemble machines like airplanes, cars, and turbines!

"It's fun to make something that works!" says Mia.
"No machine can work without us!" says Mindy.
"That's right!" says Wendy the welder.
"Together we can make anything!"

Wendy uses welders of all sorts to join metal parts and pieces. Can you think of something that's welded?

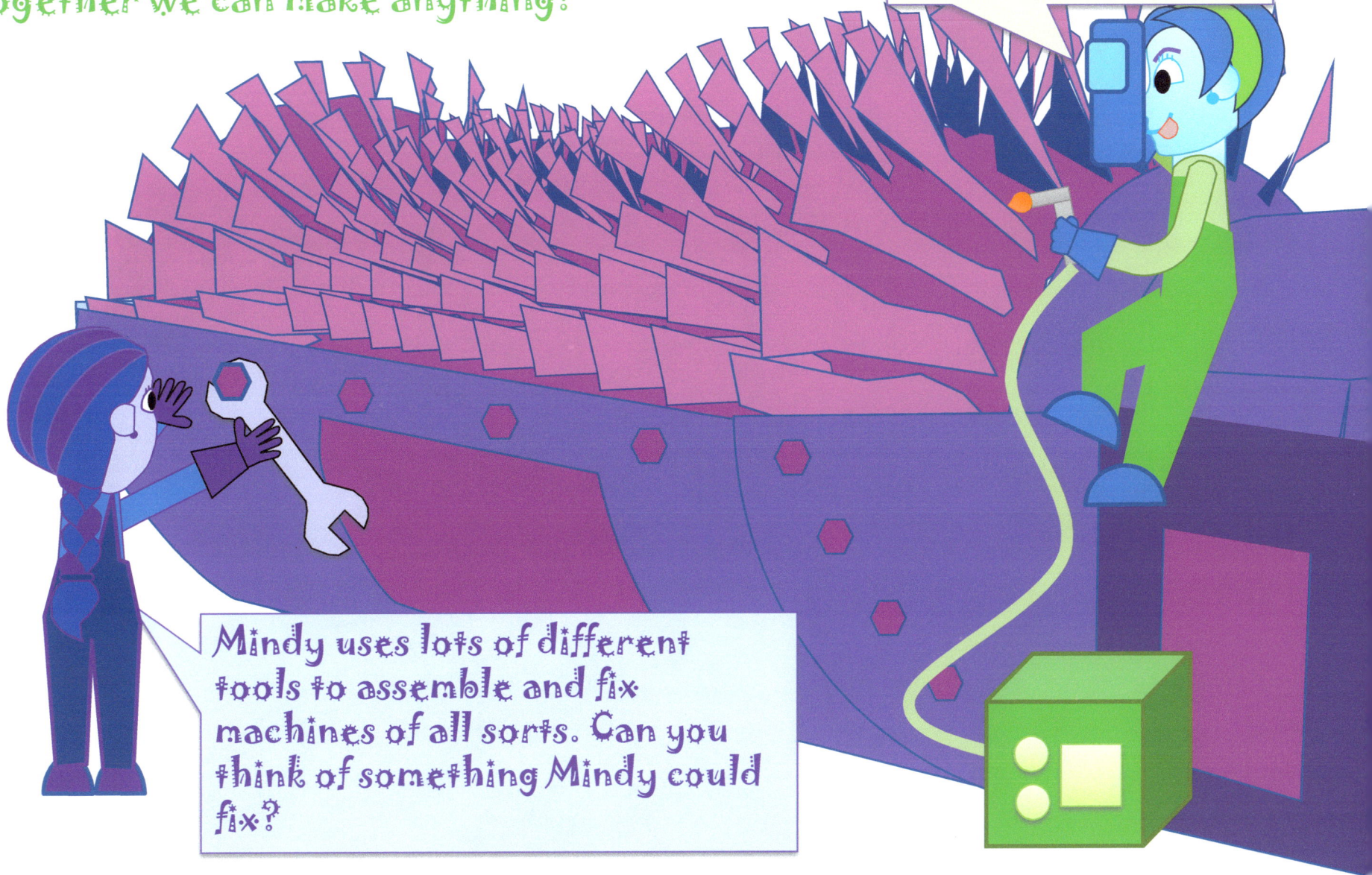

Mindy uses lots of different tools to assemble and fix machines of all sorts. Can you think of something Mindy could fix?

So that's how most electricity is made - in powerplants like you see here! Steam plants must be by the river or ocean (they need a lot of water!) Gas plants can be almost anywhere.

Steam turbines like this need lots of water! That's why coal and nuclear power plants are always by a river or ocean.

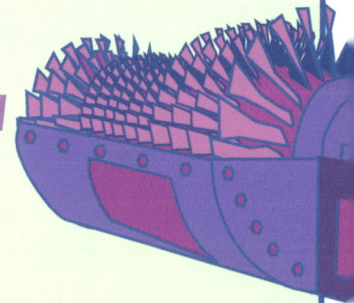

## Steam Power Plant

But there are other ways to make electricity. These are called "renewable" because they don't use anything up!

"Hydro" Power is a renewable way to make electricity because it uses water that flows by gravity (something that doesn't run out!) The water flows from a higher lake or river through a dam, and you guessed it, there are turbines spinning inside that generate the electricity!

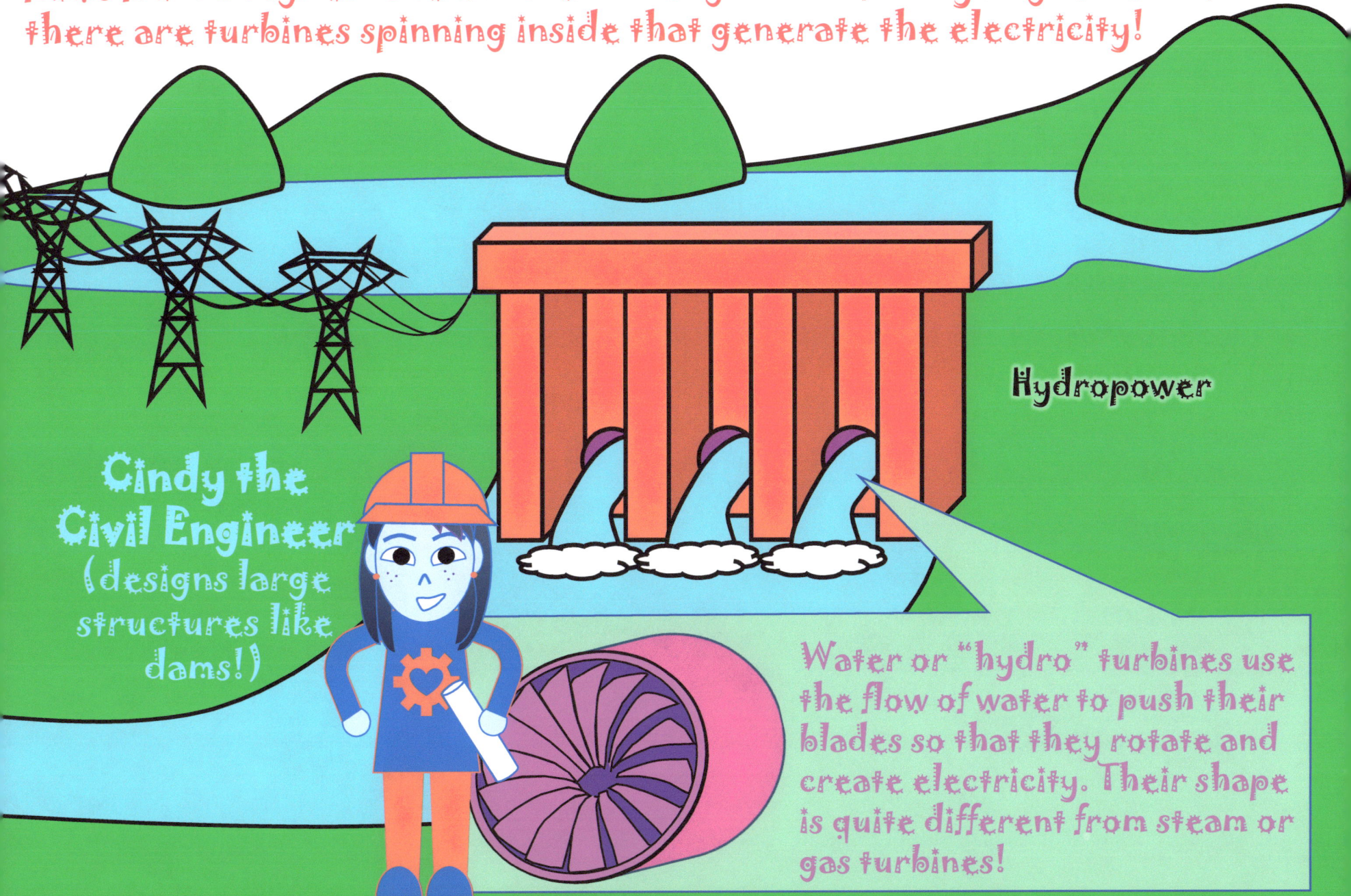

Hydropower

Cindy the Civil Engineer (designs large structures like dams!)

Water or "hydro" turbines use the flow of water to push their blades so that they rotate and create electricity. Their shape is quite different from steam or gas turbines!

Other clever scientists have discovered that energy can come from plants, too. Crops like corn can be harvested and "brewed" in a big pot to make what are called "biofuels." These biofuels can be burned just like other fuels to generate electricity.

Biofuels can be made from any *6 "organic" or living thing.

Wind power uses big fans that are pushed by, can you guess? The wind!

"Wind Turbines" work the same as gas or steam turbines except there is no heat needed! The wind blows, the blades spin round and round, pushing magnets to generate electricity.

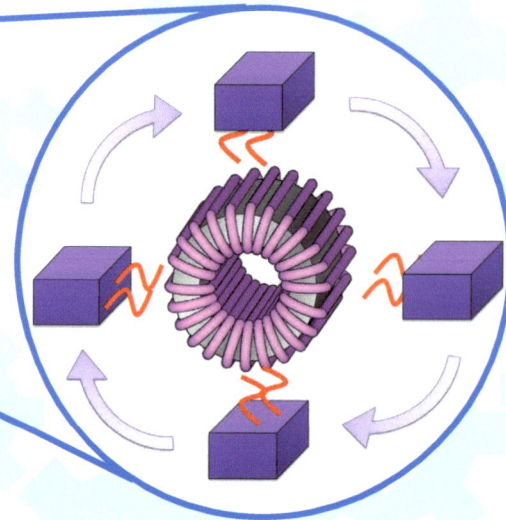

Just like all the electricity-making machines, wind turbines run because of STEMT professionals. Engineers and scientists design the machines, and trades folks make them!

"Solar energy" is a renewable energy that comes straight from the sun! It's way different than what we've seen so far. Solar panels capture the sun's light energy to make electricity. Electrical engineers like Eva help design solar panels, and trades-women like Elisa wire them up!

Eva here! I love designing electrical things like circuits, cell phones, and solar panels!

Eva the Electrical Engineer

Solar Panels

Hi! I'm Elisa. Electricians like me install the wiring and electronics that move all our electricity around!

Elisa the Electrician

So, you see, there's so much STEMT in making electricity! In fact, STEMT is what makes electricity possible!

Science, technology, engineering, math, and trades are all equally important.

And all STEMT workers should be our celebrities! After all, we wouldn't have electricity or other technologies without them.

You've just seen lots of new stuff that most people don't know! Go you! It's a lot of information, so don't worry if you have to look anything up or ask questions.

You can re-read the book to help you understand STEMT and electricity better. Sometimes it takes a few times to understand something! But that's ok, it's so fun!

I'm so glad you joined me learning about STEMT and how electricity is made. Keep learning! You can have STEMT superpowers too!

The End

But keep going to see real life **STEMT** role models! →

Meet the women in STEMT that inspired the Amelia Gearheart Series!
Visit ifthencollection.org and AmeliaGearheart.com

## Amelia Gearheart
## Mechanical Engineer

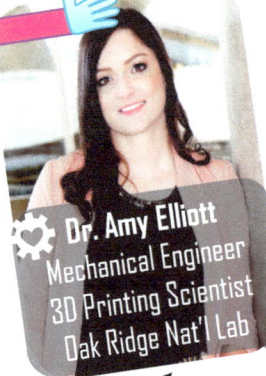

**Dr. Amy Elliott**
Mechanical Engineer
3D Printing Scientist
Oak Ridge Nat'l Lab

**Debbie Sterling**
Mechanical Engineer
& Toy Maker

Meet the Author:
Amelia ("#AmytheEngineer") Elliott

## Mimi the
## Metallurgist

**Dr. Dawn White**
Oak Ridge Nat'l Lab
Material Scientist &
Entrepreneur

**Beth Armstrong**
Materials Researcher
Oak Ridge Nat'l Lab

## Fahima The
## Physicist

**BURÇIN MUTLU-PAKDIL**
TUCSON, AZ
ASTROPHYSICIST

**Dr. Bianca Haberl**
Oak Ridge National Lab
Physicist

**Dr. Sarah Cousineau**
Oak Ridge National Lab
Physicist

**Dr. Jessica Esquivel**
Fermilab
Physicist

**DEBORAH BEREBICHEZ**
NEW YORK, NY
PHYSICIST AND DATA SCIENTIST

**Crystal Emery**
Producer of Media for
Diversity and Inclusion

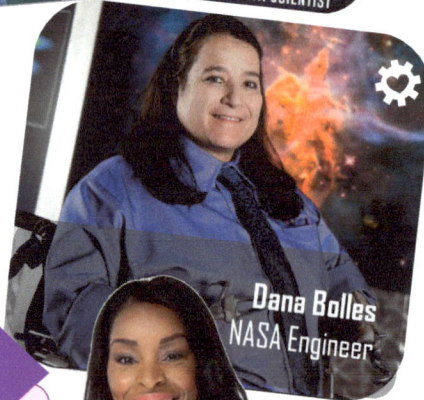

## Nina the
## Nuclear
## Engineer

**Dana Bolles**
NASA Engineer

**J'TIA HART**
ARGONNE, IL
NUCLEAR ENGINEER

(This is the Author's Twin Sister!)

**Beata Mierzwa**
La Jolla, CA
Molecular Biologist & Artist

**Manda Masino**
Austin, TX
Biologist, Professor & Research Director

**Lori Beth Browning**
Food Safety – Master's in Epidemiology

**Keisha the Chemical Engineer**

**Janis Louie**
Salt Lake City, UT
Professor of Chemistry

**Kris Inman**
Ennis, MT
Wildlife Biologist

**Danielle Twum**
San Francisco, CA
Cancer Immunologist & Translational Scientist Liaison

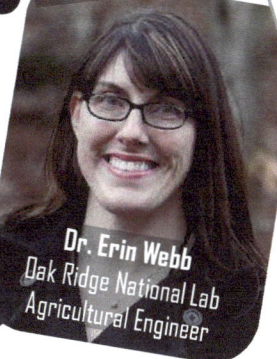

**Dr. Erin Webb**
Oak Ridge National Lab
Agricultural Engineer

**Beth the Biologist**

**Dr. Merlin Theodore**
Oak Ridge National Lab
Chemical Engineer and Material Scientist

**Dr. Santa Jansone Papova**
Oak Ridge National Lab
Organic Chemist

**Chloe Lerin**
Engine Researcher & Motorcycle Racer

**Nikki Sereika**
Aircraft Maintenance Technician

**Dr. Eva Hakansson**
Builds Electric Motorcycles
Land Speed Motorcycle Racer

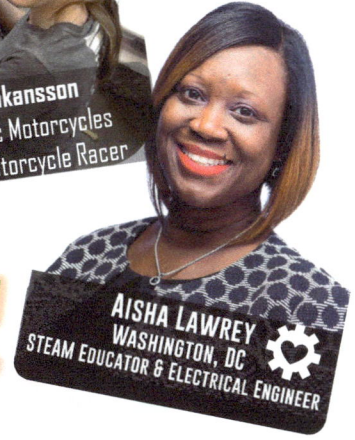

**Mia the Machinist**

**Mindy the Mechanic**

**Eva the Electrical Engineer**

**Aisha Lawrey**
Washington, DC
STEAM Educator & Electrical Engineer